ANIMAL FAMILIES

Colin Threadgall

Dragonfly Books™

Crown Publishers, Inc., New York

Pack of wolves

Wolf

Gaggle of geese

Goose

Pod of whales

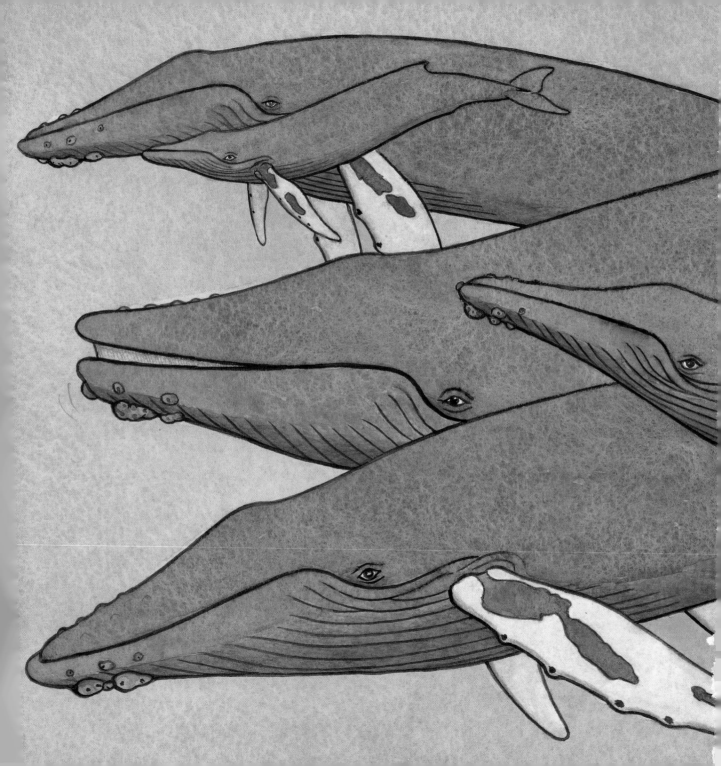

Whale

Bird

Flock of birds

Cow

Herd of cows

Pride of lions

Lion

School of fish

Fish

Dragonfly Books™ published by Crown Publishers, Inc.

Published by Crown Publishers, Inc., a Random House company, 201 East 50th Street, New York, New York 10022. Originally published in Great Britain in 1996 by Julia MacRae, Random House UK Ltd.

CROWN is a trademark of Crown Publishers, Inc.

Printed in Hong Kong

Library of Congress Cataloging-in-Publication Data
Threadgall, Colin.
Animal Families / Colin Threadgall. — 1st American ed.
p.   cm.
Summary: Labeled illustrations present the appropriate names for groups of different animals, including a gaggle of geese, a herd of cows, and a pride of lions.
1. Animals—Juvenile literature. 2. English language—Collective nouns—Juvenile literature.
[1. Zoology—Nomenclature. 2. English language—Collective nouns.] I. Title.
QL49.T38   1996
591—dc20   95-32825

ISBN 0-517-88548-4

10 9 8 7 6 5 4 3 2 1

First American Edition